THE ANIMAL BIOGRAPHY SERIES

NEFERTITI
THE SPIDERNAUT

Darcy Pattison
WRITTEN BY

Valeria Tisnés
ILLUSTRATED BY

The Jumping Spider
Who Learned to Hunt in Space

Nefertiti, the Spidernaut: The Jumping Spider Who Learned to Hunt in Space
by Darcy Pattison
Illustrated by Valeria Tisnés

Thanks to the followng for their expert help: Stefanie Countryman at BioServe, Boulder, CO; Paula E. Cushing, Ph.D., Curator of Invertebrate Zoology, Denver Museum of Nature & Science; and, Astronaut Sunita Williams, Captain U.S.Navy for their expert help.
For H and B. Thanks for your help!

Mims House
1309 Broadway
Little Rock, AR 72202

Publisher's Cataloging-in-Publication data

Names: Pattison, Darcy, author. | Tisnés, Valeria, illustrator.
Title: Nefertiti, the spidernaut: the jumping spider who learned to hunt in space / By Darcy Pattison; Illustrated by Valeria Tisnés.
Description: Little Rock, Arkansas: Mim's House, 2016.
Identifiers: ISBN 9781629440606 (Hardcover) | 9781629440613 (pbk.) | 9781629440620 (ebook) | LCCN 2015920985.
Summary: A jumping spider is sent to the International Space Station to discover if she can hunt in microgravity.
Subjects: LCSH Space biology–Juvenile literature. | Animal experimentation –Juvenile literature. | International Space Station–Research–Juvenile literature. | Space stations–Juvenile literature. | Space flight–Physiological effect–Juvenile literature. | Jumping spiders–Juvenile literature. | BISAC JUVENILE NONFICTION / Animals / Insects, Spiders, etc. | JUVENILE NONFICTION / Technology / Aeronautics, Astronautics & Space Science.
Classification: LCC QH327 .P38 2016 | DDC 629.45009–dc23

Early in 2012, in a greenhouse near Yarnell, Arizona, a Johnson jumping spider laid an egg sac.

About the same time, Stefanie Countryman was looking for a jumping spider to send on a space mission. Scientists hypothesized, or predicted, that jumping spiders would not be able to hunt in the microgravity of the International Space Station. On Earth, a spider would jump to catch its food and then land on the ground. On the International Space Station, however, there is little gravity to pull them back down. Instead, the spider would just float.

The egg sac arrived at Stefanie's office in Boulder, Colorado on a Friday afternoon in February, 2012. The package included a bag of fruit fly larvae for the spiderlings to eat. Stefanie placed the spider egg sac and the open bag of fruit fly larvae into a covered box for the weekend. On Monday morning, Stefanie was horrified. Dozens of tiny spiderlings just the size of a pinhead had hatched.

Almost all the fruit flies had hatched, too, and the mass of flies had killed almost all the spiderlings. Only four jumping spiders were still alive for Stefanie to raise.

One day, Stefanie had to choose a spider for the space mission. She chose the largest female spider because her red abdomen was easy to see. She was also the most active spider. They named her Nefertiti after an ancient Egyptian queen. Before the experiment could fly on the International Space Station, the spider had to be tested in her special habitat, or home. The spider's home measured 6" wide x 5" high x 3" deep, about the size of a man's hand. The spider habitat had a chamber for water.

Other chambers held fruit flies, which Nefertiti would hunt and eat for her 100 days in space.

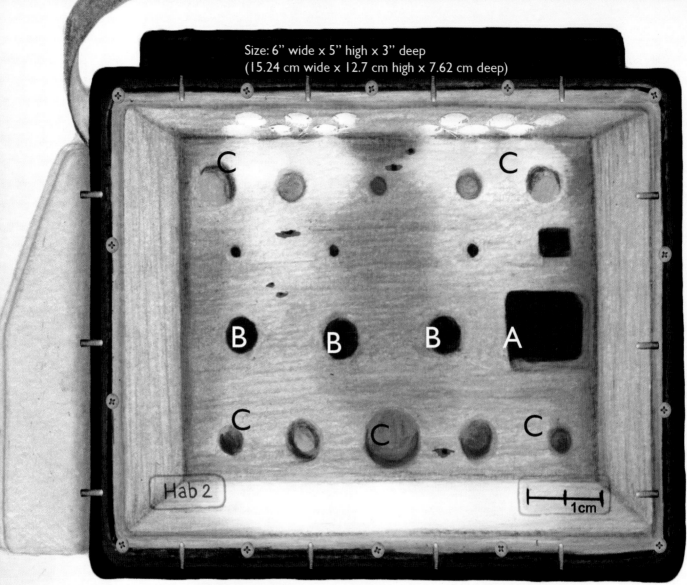

Size: 6" wide x 5" high x 3" deep
(15.24 cm wide x 12.7 cm high x 7.62 cm deep)

Hab 2

1cm

Spider's Habitat (approximately life size)
A. Water Chamber
B. Fruit Fly Chambers
C. Indented Areas for Spider to Hide

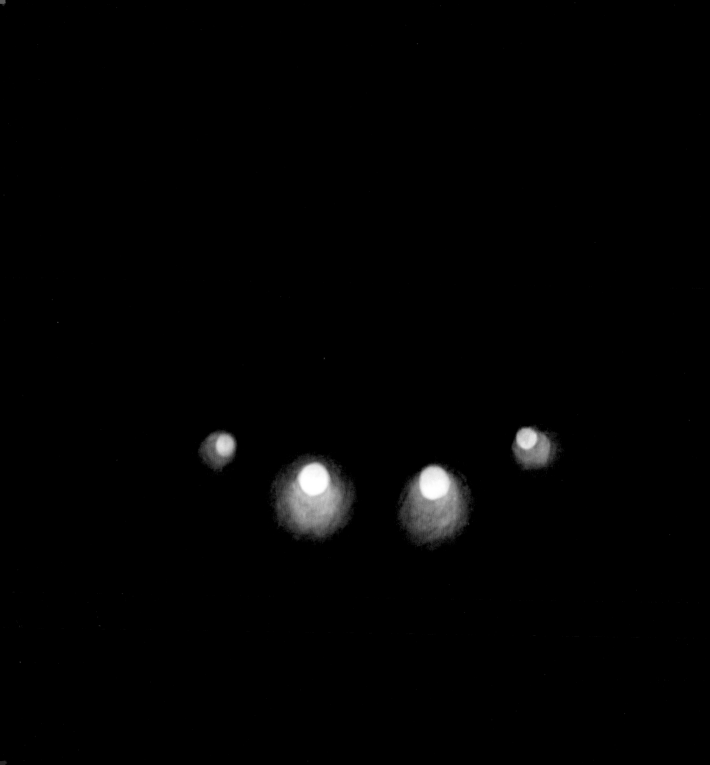

For the first survival test, Nefertiti lived for a week with no light. That was hard because she only hunted by day. Later, she spent a week in extreme cold. The light and cold tests didn't bother Nefertiti, so scientists agreed: Nefertiti would be the spidernaut.

In July, Stefanie fed Nefertiti until she was gorged and her abdomen was big. She wouldn't eat again until she reached the space station.

Stefanie took Nefertiti in her habitat and flew from Colorado to Japan, and then they traveled to the island of Tanegashima. There, on July 17, Nefertiti was handed over to the Japanese Space Agency. They packed Nefertiti and her equipment first into container #1094, and then onto the HTV-3 unmanned cargo carrier.
Rain poured on July 21, but the HTV-3 launched anyway. Stefanie wondered: would the spidernaut return to Earth alive?

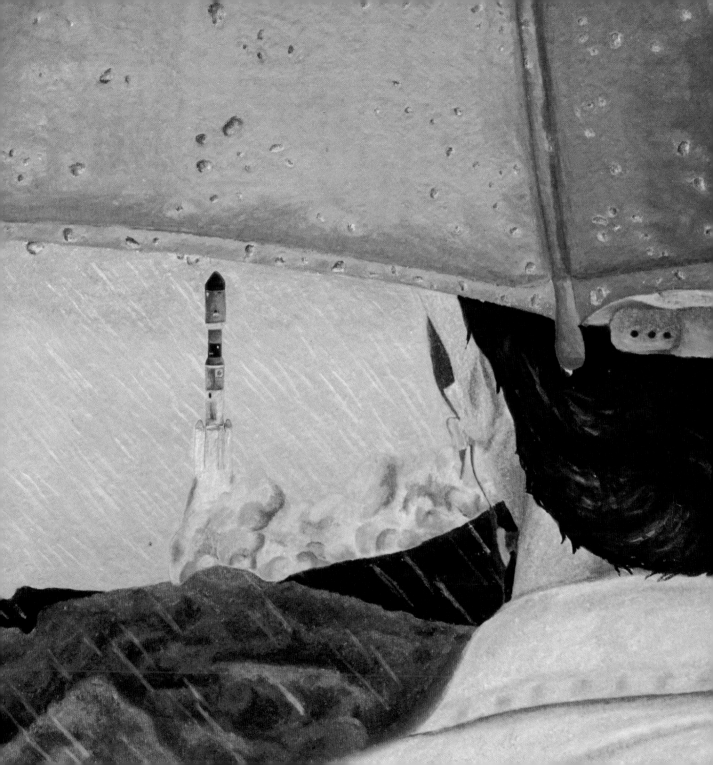

Six days later, the HTV-3 berthed, or attached, onto the International Space Station. The spider experiment was unpacked. Nefertiti got her first look at Sunita (Suni) Williams, (Captain, U.S. Navy), the astronaut who would care for her. First, Suni unpacked and set up the habitat. Nefertiti and the fruit flies floated about.

As usual,
Nefertiti
spun a silk line
and attached it to the wall as a tether.
She leapt for a fruit fly. On Earth, Nefertiti jumped
in an arc and almost never missed her prey.
But in the
microgravity
of the
International
Space Station, she hung almost
weightless in mid-air, just as
the scientists had predicted.

Hungry,
she tried again.
And missed.

For three hours a day, Stefanie watched video feed from the space station and worried. Clinging to the balsa wood walls, Nefertiti ran around looking for food.

On July 24, Nefertiti stalked a fruit fly.
This time, instead of leaping in an arc,
she lunged in a straight line.
She caught the fruit fly!

The hypothesis was wrong. The Johnson jumping spider did adapt to microgravity. Nefertiti learned to jump in a new way to catch her food.

Even more surprising, a few days later, Stefanie saw Nefertiti hunt again. This time, she anchored her dragline and leapt for a fruit fly in mid-air. She caught the fly, and like a bungee cord, her dragline pulled her back to the habitat's wall.

Nefertiti was hunting as no other jumping spider had ever hunted before.

After two weeks of videotaping, the experiment was officially over. However, it would be eleven more weeks before a space capsule could return Nefertiti to Earth. Suni kept Nefertiti's habitat near her work desk. She often stopped to watch the spider hunt and eat. And the spider often stopped to watch Suni moving around.

Sometime late in September, Nefertiti caught and ate the last fruit fly in her habitat. Now, she lived on just water. Meanwhile, the International Space Station flew around and around Earth. In all, Nefertiti circled Earth about 1584 times.

Over a month later, on October 27, Suni packed Nefertiti and her habitat for the return trip to Earth. Suni wrote in her NASA blog, "... hang in there, there are many fruit flies on Earth..." *

The Dragon capsule splashed down on October 28 in the Pacific Ocean near Long Beach, California. When Stefanie opened the habitat, Nefertiti was alive, but barely. Somehow during the return trip, her water compartment had been closed off. Immediately, Stefanie gave Nefertiti a wet cotton ball, and the spider drank and drank.

Back in Colorado, Nefertiti had one last test. Would she be able to re-adapt to Earth's gravity and hunt? She'd lived for over forty days with no food. Her abdomen was almost flat.

A fruit fly buzzed around the spider's habitat. Nefertiti anchored a dragline and sprang. She overshot the fly and slammed into the front window of the habitat. She tumbled to the floor onto her back.

She righted herself and lunged again for the fruit fly. She missed. Again, she pounced. And missed. Desperately hungry, Nefertiti stalked the fruit fly. She crept forward, one step, two steps—and then, she leapt. And ate.

Nefertiti had relearned
how to hunt on Earth.

For several weeks, Stefanie fattened up the spidernaut with small crickets. Nine-month-old Nefertiti finally had enough to eat.

But a famous spidernaut couldn't live
on Stefanie's kitchen counter.
On November 29, Nefertiti flew again.
This time she traveled to Washington D.C.,
where she would live in the
O. Orkin Insect Zoo
at the Smithsonian
National Museum of
Natural History.

Sadly, on
December 3,
keepers found
Nefertiti the
Spidernaut
dead
of old age.

In her ten-month life, Nefertiti traveled over 42 million miles. She hunted on Earth and surprised scientists by hunting successfully in space. Her life and journey built excitement for the space program. And perhaps, along the way, she taught some people to appreciate the simple beauty of a spider.

FACTS ABOUT NEFERTITI

Johnson Jumping Spider, *Phiddipus johnsonii*

Spiders are wingless, have two body parts and eight legs. A spider's exoskeleton is the hard outer shell of its body. Spiders are carnivorous, which means they eat other animals, including insects. Some spiders spin a web and wait for something to get caught. Jumping spiders do produce silk, but they shape it like a tube and use it for shelter or laying eggs. To find food, jumping spiders are active hunters. Their excellent eyesight helps them hunt. Before jumping to catch something to eat, the spider usually lays down a dragline as an anchor thread. Jumping insects often have enlarged and muscular legs. However, arachnids like the Johnson jumping spider rely on hydraulic pressure, or the rapid inflating of the legs. Leg muscles contract, which forces liquid into the hind legs, which makes the spider extend the legs quickly in a forceful jump. The Johnson jumping spider crushes prey with powerful jaws, or chelicerae, and injects venom through its fangs. Then, the spider regurgitates, or throws up, digestive juices called enzymes, which turns prey into a liquid for the spider to slurp up.

LIFESPAN: Expected lifespan is about a year. Nefertiti lived mid-February - December 3, 2012, about 10 months.

RANGE: Johnson jumping spiders are common in western North America, including Mexico, United States, and Canada.

LIFE CYCLE: Eggs are laid, or oviposited, in batches and hatch three weeks later. The spiders molt, or shed their exoskeleton, 5-8 times until sexually mature; they may continue to grow larger in size with increased food. Females are larger than males and have a black stripe down their red abdomen.

LENGTH: About 1/2" (1-1.5 cm) when full-grown.

VOCABULARY

SPIDERNAUT: Spidernaut is a made-up word. The first part tells you that this is about a spider. The second part comes from the last half of the word astroNAUT. By combining the words, we understand that a spidernaut is a spider who went to space.

MICROGRAVITY: Space is a weightless environment, but the International Space Station is in a low Earth orbit at about 200 to 250 miles high. At that distance, it's a microgravity environment. Gravity is the force that pulls people toward something, like the Earth. "Micro" is Latin for small, so microgravity means a place where the pull of gravity is small or weak. In microgravity, objects are almost weightless, but not quite. For more see bit.ly/NASAMicrogravity

HYPOTHESIS: A hypothesis is a statement about what scientists think will or will not happen as the result of an experiment. The experiment proves or disproves the hypothesis.

READ AND WATCH MORE:

- Sunita Williams, the International Space Station astronaut in charge of the spider experiment, blogged daily about her duties. On August 10, 2012, she wrote about the spider experiment: http://bit.ly/AstroSuniBlog
- Watch this video of Nefertiti hunting in space (0:00 – 0:15) and then readapting to Earth's gravity (0:15 – 0:57): bit.ly/NefertitiHunting

*Blog quote on p. 22 Astro Suni's Blog by Sunita Williams.https://astrosuni.wordpress.com/2012/10/. Retrieved September 2015.

ENGINEERING PROBLEM: SPIDER HABITAT FOR SPACE

THE PROJECT: In October, 2011, YouTube Space Lab announced a competition for students ages 14-18. They asked students to submit a video explaining a science experiment they'd like to see sent to the International Space Station (ISS). The competition was sponsored by YouTube and Lenovo, and conducted in collaboration with Space Adventures, NASA, European Space Agency and Japan Aerospace Exploration Agency. One of the two projects chosen was proposed by 18-year-old Amr Mohamed of Alexandria, Egypt. (Learn more about Amr in this video: http://bit.ly/MeetAmr) Amr wondered what would happen when a spider jumped in a microgravity environment. The jumping spider experiment was transformed into a successful space flight investigation by Stefanie Countryman and others at BioServe Space Technologies (http://www.colorado.edu/engineering/BioServe/), a center at the University of Colorado that specializes in creating space flight habitats that enable living organisms to exist as naturally as possible in an unnatural environment. Amr named the spiders Cleopatra and Nefertiti, in honor of queens of ancient Egypt. Cleopatra, a zebra spider (*Salticus scenicus*), rarely came out when the video camera was filming, so Nefertiti was considered the main spider in the experiment.

THE HABITAT: BioServe Space Technologies has sent sixteen spiders into space since 1973. The original spider habitat was a 6" wide x 5" high x 3" deep box was made of light-weight plastic and anodized aluminum. The interior was lined with a narrow frame of light-weight balsa wood. This basic design met the mass and volume requirements for an ISS experiment. They tried to improve that design for Nefertiti.

One big engineering and biological challenge was how to feed a spider in space. Fruit flies (*Drosophila*) are an easy source of food because they can live in microgravity; however, they only live 40-50 days. Spiders can live for long periods with only water, but engineers still looked for ways to provide food for the full 100 days of Nefertiti's flight. For Nefertiti's habitat, scientists and engineers decided to try to raise several generations of fruit flies. Engineers created a mini-hab with chambers that attached to the back of the habitat. Chamber 1 contained water for the spider. Chamber 2 contained the original fly larvae. When the habitat reached the ISS, the astronaut opened Chamber 2 to release the newly hatched fruit flies. She also opened Chamber 3, which held more fruit fly food flakes. The engineers hoped the fruit flies would mate and lay eggs in Chamber 3. That would produce a second generation of fruit flies. After a couple of weeks, astronauts were instructed to open Chamber 4. They hoped the fruit flies would again lay eggs, creating a third generation of flies. If all three generations worked, they'd have enough food to feed the spider for 60-70 days of flight, but not the full 100 days of flight.

The habitat was considered a success. In the end, Nefertiti had food for about 60 days. Her natural ability to survive on just water kept her alive the last 40 days. The next time BioServe sends spiders into space, they'll improve the design of the habitat. Often scientists and engineers can't solve all the problems at one time. Instead, they make a small change and test it. If that works, they make another small change and test that. Eventually these small changes add up to big changes and a successful design. This type of "incremental changes" in an experiment is part of the engineering and technology that went into the design of the spider habitat for the ISS project. For more, see this video: Bioserve Space Techonology explains the habitats available for use on the ISS: http://bit.ly/SpiderHabitat